ESSENTIALS

C000192794

CONTENTS

04 [CHAPTER ONE]
GET STARTED WITH GPIO ZERO
Discover what it's all about

11 [CHAPTER TWO]
CONTROL LEDS
Make them blink and more

16 [CHAPTER THREE]
ADD A PUSH BUTTON
Create a fun reaction game

22 [CHAPTER FOUR]
MAKE A MUSIC BOX
Link buttons to sounds

30 [CHAPTER FIVE]
LIGHT AN RGB LED
All the colours of the rainbow

36 [CHAPTER SIX]
MOTION-SENSING ALARM
Use a PIR to create an intruder alert

40 [CHAPTER SEVEN]
BUILD A RANGEFINDER
Connect and read a distance sensor

45 [CHAPTER EIGHT]
MAKE A LASER TRIPWIRE
Use an LDR to detect light/dark

51 [CHAPTER NINE]
BUILD AN INTERNET RADIO
Adjust the station with potentiometers

59 [CHAPTER TEN]
CREATE AN LED THERMOMETER
Read an analogue temperature sensor

66 [CHAPTER ELEVEN]
BUILD A ROBOT
Control DC motors with GPIO Zero

74 [CHAPTER TWELVE]
QUICK REFERENCE
A handy guide to GPIO Zero's many useful classes

[PHIL KING]

Phil King is a Raspberry Pi enthusiast and regular contributor to *The MagPi* magazine. Growing up in the 'golden era' of 8-bit computers in the 1980s, he leapt at the chance to write about them in magazines such as *CRASH* and *ZZAP!64*. When consoles took over the video games world, he missed the opportunity to program... until the Raspberry Pi came along. Phil is now an avid coder and electronics dabbler, who loves to work on projects with his six-year-old son.

The MagPi ESSENTIALS

[CHAPTER ONE]
GET STARTED WITH
ELECTRONICS
& GPIO ZERO

Discover what GPIO Zero is and how you can use it to
program electronics connected to your Raspberry Pi

[COMMON
COMPONENT
SYMBOLS]

>RESISTOR

>BUTTON/
SWITCH

>LED

>CAPACITOR

T he Raspberry Pi is great for learning computing. Whether that's coding or tricks for the advanced user, the Raspberry Pi has many tools to help you learn about them. It's also very good at physical computing, which in this context means programming and interacting with the real world through electronics. In simple terms, physical computing with the Pi is something like programming it to turn on an LED, a component in an electronic circuit.

The electronic circuits are the physical part of a physical computing project connected to the Raspberry Pi. These circuits can be simple or very complex, and are made up of electronic components such as LEDs, buzzers, buttons, resistors, capacitors, and even integrated circuit (IC) chips.

At its simplest, an electronic circuit lets you route electricity to certain components in a specific order, from the positive end of a circuit to the negative (or ground) end. Think of a light in your house: the electricity passes through it, so it lights up. You can add a switch that breaks the circuit, so it only lights up when you press the switch. That is an electronic circuit.

Reading circuit diagrams

Building a circuit can be easy if you know what you're doing, but if you're making a new circuit or are new to electronics in general, you'll most likely have to refer to a circuit diagram. This is a common way you'll see a circuit represented, and these diagrams are much easier to read and understand than a photo of a circuit. However, components are represented with symbols which you'll need to learn or look up. **Fig 1** is an example of the light circuit we talked about before. Here we

Fig 1 Switch circuit

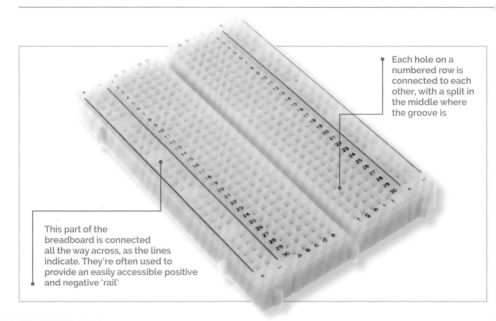

Each hole on a numbered row is connected to each other, with a split in the middle where the groove is

This part of the breadboard is connected all the way across, as the lines indicate. They're often used to provide an easily accessible positive and negative 'rail'

have a power source (a battery in this circuit), a switch, a resistor, and an LED. The lines represent how the circuits are connected together, either via wire or other means. Some components can be used any way round, such as the resistor or switch. However, others have a specific orientation, such as the LED. Diodes only let electricity flow from positive to negative; luckily, real-life LEDs have markers such as longer legs or a flat edge to indicate which side is positive, making them easier to wire up.

The Raspberry Pi and electronic circuits

Making a Raspberry Pi part of the circuit is quite easy. At its most basic, it can provide power to a circuit, as well as a negative or ground end through the GPIO pins. Some pins are specifically always powered, mostly by 3.3V, and always go to ground. Most of them can be programmed to create or recognise a HIGH or LOW signal, though; in the case of the Raspberry Pi, a HIGH signal is 3.3V and a LOW signal is ground or 0V.

In an LED example, you can wire up an LED directly to a 3.3V pin and a ground pin and it will turn on. If you instead put the positive end of the LED onto a programmable GPIO pin, you can have it turn on by making that pin go to HIGH (see chapter 2 for more details).

Wiring up a circuit to a Raspberry Pi is fairly simple. To create the physical circuits in the guides throughout this book, we're using prototyping breadboards. These allow you to insert components and wires to connect them all together, without having to fix them permanently. You can modify and completely reuse your components because of this.

[GPIO NUMBERS]

40-pin GPIO header key for Raspberry Pi 3, 2, B+, and A+

3.3V	1	2	5V
GPIO2	3	4	5V
GPIO3	5	6	GND
GPIO4	7	8	GPIO14
GND	9	10	GPIO15
GPIO17	11	12	GPIO18
GPIO27	13	14	GND
GPIO22	15	16	GPIO23
3.3V	17	18	GPIO24
GPIO10	19	20	GND
GPIO9	21	22	GPIO25
GPIO11	23	24	GPIO8
GND	25	26	GPIO7
DNC	27	28	DNC
GPIO5	29	30	GND
GPIO6	31	32	GPIO12
GPIO13	33	34	GND
GPI19	35	36	GPIO16
GPI26	37	38	GPIO20
GND	39	40	GPIO21

GPIO Zero uses the BCM GPIO numbering rather than the pin number – use this handy table to remember which is which

Using GPIO Zero

Once the components are all hooked up to your Raspberry Pi, you need to be able to control them. The Raspberry Pi is set up to allow you to program it with the Python language. This has also been made simpler recently with the addition of GPIO Zero. It comes pre-installed in the latest version of Raspbian Jessie. If you don't have it yet, however, you can install GPIO Zero manually: after performing a package list update by entering **sudo apt-get update** in a terminal, enter **sudo apt-get install python3-gpiozero**.

GPIO Zero was created to simplify the process of physical computing, helping new coders to learn. It's a Python library which builds upon the existing GPIO libraries RPi.GPIO, RPIO, and pigpio. However, while those libraries provide an interface to the GPIO pins themselves, GPIO Zero sits above them and provides a way to interface to the devices that you connect to those pins.

This change simplifies thinking about physical computing. Consider wiring a simple push button to GPIO 4 and ground pins. In order to react to this button, we need to know that the pin should be configured with a pull up resistor, and that the pin state when the button is pushed will be 0. Here's what this would look like in the classic RPi.GPIO library:

```
from RPi import GPIO

GPIO.setmode(GPIO.BCM)
GPIO.setwarnings(False)
GPIO.setup(4, GPIO.IN, GPIO.PUD_UP)
while GPIO.input(4):
    pass
print("Button pushed!")
```

To complete beginners, there's quite a lot going on there, which gets in the way of actually experimenting with it and even teaching the simple logic required. Here's the equivalent code in GPIO Zero:

```
from gpiozero import Button

btn = Button(4)
while not btn.is_pressed:
    pass
print("Button pushed!")
```

The boilerplate, the code you have to blindly enter without understanding why you're entering it, is reduced to the bare minimum that we need. The name 'GPIO Zero' derives from this 'zero boilerplate' philosophy, which was first espoused by Daniel Pope's Pygame Zero library. The logic is also straightforward, with no curious inversion of the input value. So, now you've learnt about GPIO Zero and how it makes coding much simpler, it's time to get started doing some physical computing with it. In chapter 2, we'll show you how to wire up some LEDs on a breadboard and control them using GPIO Zero's **LED** class.

[WHAT'S NEW IN GPIO ZERO?]

GPIO Zero can use buttons, LEDs, buzzers, and lots of other components. The library is always expanding.

>SERIAL PERIPHERAL INTERFACE

Released in 1.2.0, there is now an SPI implementation for specific compatible devices to talk to the Pi. This allows for analogue inputs, analogue-to-digital converters, and other pretty advanced stuff, but it makes using them much more simple.

>HOLD EVENTS

These are variables in something like button code that allow you to set a length of time a button should be pressed before being recognised as a press. This can be useful if your button is very twitchy in a project you're using.

>SOURCE TOOLS

The tools library for the **source** and **values** properties enables you to tweak and play with the way GPIO Zero handles specific components and functions. We won't be covering them this issue, but they're important for advanced projects.

[CHAPTER TWO]
CONTROL LEDS
WITH GPIO ZERO

Turn LEDs on and off with just a few lines
of code, and build a traffic light system

You'll Need

> GPIO Zero

> 1× solderless breadboard

> 3× LEDs (red, yellow, green)

> 3× 330Ω resistors

> 4× male-to-female jumper wires

> Or a Traffic HAT **magpi.cc/ 1Mma7oD**

A resistor is required to limit the amount of current being drawn by the LED, to avoid damage to the Pi

The LED's longer leg is wired to GPIO 25, while the other is connected via a resistor to the ground rail

---

One of the first physical computing projects you'll want to try with GPIO Zero is lighting an LED. This is very simple to achieve using the library's **LED** class, using very few lines of code. Here we'll show you how to wire up a simple circuit connected to your Raspberry Pi's GPIO pins, then light an LED and make it blink on and off. We'll then add two more LEDs to make a traffic light system, or you can also use a special Traffic HAT add-on.

>STEP-01
Connect an LED

It's best to turn the Pi off when building a circuit. The breadboard features numbered columns, each comprising five connected holes. Place your red LED's legs in adjacent numbered columns, as shown in the diagram. Note that the shorter leg of the LED is the negative end; in its column, insert one end of the resistor, then place the other end in the outer row marked '−' (the ground rail). Use a male-to-female jumper wire to connect another hole in that ground rail to a GND pin on the Pi. Finally, use a jumper wire to connect a hole in the column of the LED's longer (positive) leg to GPIO pin 25.

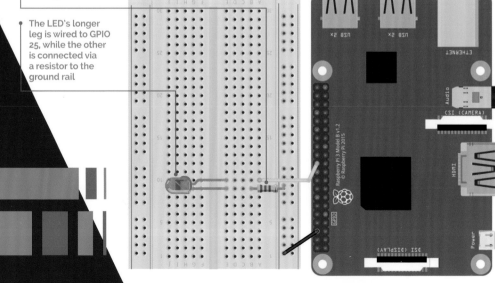

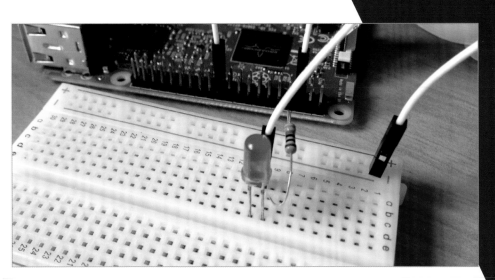

>STEP-02
Light the LED

We'll now test our circuit with a simple Python program to make the LED turn on and off. To start coding, open IDLE from the Main Menu: Menu > Programming > Python 3 (IDLE). Create a new file by clicking File > New file. Save it with File > Save, naming it **ch2listing1.py**. Now enter the code from the listing of the same name (page 16).

At the start of the program, we import the **LED** class from GPIO Zero, and the **sleep** function from the **time** library (to enable us to pause between turning the LED on and off). We then assign the **led** variable to the GPIO 25 pin, which will power it whenever we set it to on in the code. Finally, we use **while True:** to create a never-ending loop that switches the LED on and off, pausing for 1 second between each change. Press **F5** to run the code, and your LED should be flashing on and off. To exit the program, press **CTRL+C**.

Above While it's possible to connect an LED and resistor directly to the Pi, it's better to use a solderless breadboard

>STEP-03
Easier blinking

Alternatively, to make things even easier, GPIO Zero features a special **blink** method. You could try entering the code from **ch2listing2.py** (page 16), which does exactly the same thing as the first listing, but with even fewer lines of code.

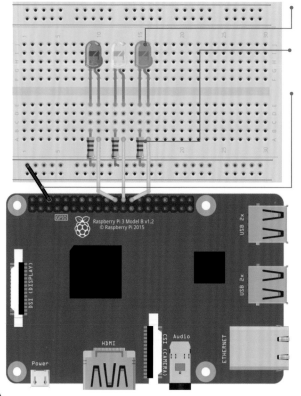

Each LED's longer leg is wired to the respective GPIO pin, while the other is connected via a resistor to the ground rail

A resistor is required to limit the amount of current being drawn by each LED, to avoid damage to the Pi

Each LED circuit shares a common ground via the '−' rail, which is connected to a GND pin on the Pi

Note that between the brackets for **led.blink**, you can add parameters to set the on and off times, number of blinks, and determine whether it runs as a background thread or not.

>STEP-04
Add more LEDs

Now that we've got the hang of controlling one LED, let's add a few more and create a traffic light sequence. You can add an optional push button to control it if you like, but for now we'll stick to just the LEDs. Connect them as shown in the diagram, with the longer (positive) legs connected via jumper wires to the following GPIO pins: 8 (yellow), and 7 (green). As before, we need a resistor for each LED, which shares a common ground connection via the '−' rail to one of the Pi's GND pins.

>STEP-05
Enter the code

After opening Python 3 (IDLE), type in the code from **ch2listing3.py** (page 17) and save it. As before, we import the **LED** class and **sleep** function from GPIO Zero and the **time** library respectively. We then assign **red**, **amber**, and **green** variables to the relevant GPIO pins. To start with, we turn the green LED on and the others off. Finally, we use **while True:** for a never-ending loop; this waits 10 seconds before showing amber then red, then waits another 10 seconds before showing red/amber then green. Press **F5** to run the program and wait for the traffic light sequence to start.

Rather than using **sleep** to create a delay between each sequence, you could trigger it with the addition of a push button: see chapter 3 for more details. You could also use a Traffic HAT with a built-in push button, LEDs, and buzzer.

>STEP-06
Traffic HAT

The Traffic HAT is a fun little kit and has its own GPIO Zero class for easy programming. With your Raspberry Pi turned off, slot the Traffic HAT over the GPIO pins, with the board itself lying across the Pi. Open a new file in Python 3 IDLE, enter the code from **ch2listing4.py** (page 17), and save it. At the top, we import the **TrafficHat** class, along with the **sleep** one from the **time** library. We then use a **while True:** loop to control the traffic lights. The green light is lit until the button is pressed, then the sequence is triggered; when it reaches red, the buzzer beeps 20 times, as on a pedestrian crossing. Amber then flashes, before it returns to green at the start of the loop, awaiting the next button press.

Above The Traffic HAT features LEDs, along with a button and buzzer

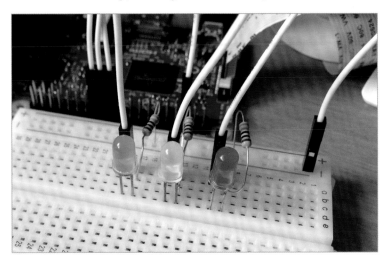

Left The three LEDs share a ground connection via their resistors, and are hooked up to GPIO pins 25, 8, and 7

ch2listing1.py

```python
from gpiozero import LED
from time import sleep

led = LED(25)

while True:
    led.on()
    sleep(1)
    led.off()
    sleep(1)
```

ch2listing2.py

```python
from gpiozero import LED
from signal import pause

red = LED(17)

red.blink()

pause()
```

ch2listing3.py

```python
from gpiozero import LED
from time import sleep

red = LED(25)
amber = LED(8)
green = LED(7)

green.on()
amber.off()
red.off()

while True:
    sleep(10)
    green.off()
    amber.on()
    sleep(1)
    amber.off()
    red.on()
    sleep(10)
    amber.on()
    sleep(1)
    green.on()
    amber.off()
    red.off()
```

ch2listing4.py

```python
from gpiozero import TrafficHat
from time import sleep

th = TrafficHat()
try:
    while True:
        # Traffic light code
        # First, turn the green LED on
        th.lights.green.on()
        print("Press the button to stop the lights!")
        # Next, we want to wait until the button is pressed
        while(th.button.is_pressed == False):
            # While not pressed, do nothing
            pass
        # Button has been pressed!
        th.lights.green.off()
        # Amber on for a couple of seconds
        th.lights.amber.on()
        sleep(2)
        th.lights.amber.off()
        # Turn the red on
        th.lights.red.on()
        # Buzz the buzzer 20 times with 0.1 second intervals
        th.buzzer.blink(0.1,0.1,20,False)
        sleep(1)
        th.lights.red.off()
        # Red off and blink amber 4 times with 0.5 second intervals
        th.lights.amber.blink(0.5,0.5,4,False)

except KeyboardInterrupt:
    exit()
```

[CHAPTER THREE]
ADD USER INPUT
WITH A PUSH
BUTTON

Make things happen at the press of a button,
and create a fun two-player reaction game

As well as output devices such as LEDs and buzzers, the Raspberry Pi's GPIO pins can be linked to input devices. One of the most basic is a simple push button, which can be used to trigger other components or functions. First, we'll hook up a button on a breadboard and get a program to print a message on the screen when it's pushed. We'll then get it to light an LED, before adding a second button for a fun two-player reaction game.

>STEP-01
Connect the button

It's advisable to turn the Pi off when building your circuit. Note: if you've already completed chapter 2, you can leave your breadboard circuit as it is, but here we'll assume you're building a new circuit.

Add the push button to the breadboard, as in the diagram, with its pins straddling the central groove. Connect a male-to-female jumper wire from one pin's column to GPIO pin 21 on the Pi. Then connect a

The LED's longer leg is wired to GPIO 25, while the other is connected via a resistor to the ground rail

When pressed, the push button pulls input pin GPIO 21 (pulled high by default) low

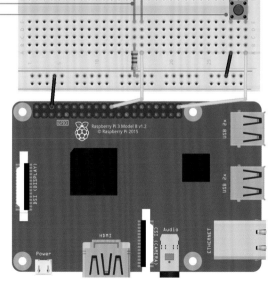

male-to-male jumper wire from the other pin (on the same side of the groove) to the '−' ground rail. Finally, connect a male-to-female jumper wire from the latter to a GND pin on the Pi.

>STEP-02
Button pushed

We'll now test our circuit with a simple Python program to show a message on the screen whenever the button is pushed. To start coding, open IDLE from the Main Menu: Menu > Programming > Python 3 (IDLE). Create a new file by clicking File > New file. Enter the code from **ch3listing1.py** (page 23), then save it.

At the start of this short program, we import the **Button** class from GPIO Zero. We then assign the **button** variable to the GPIO 21 pin, so we can read its value. Finally, we use **while True:** to create a never-ending loop that checks whether the button has been pressed or not, and prints a status message on the screen. When you run the code with **F5**, you'll get a scrolling list of messages that change when you press the button. To exit the program, press **CTRL+C**.

Note that it's also possible to trigger a Python function when the button is pressed, using the following syntax:

```
button.when_pressed = functionname
```

>STEP-03
Wait for it

GPIO Zero's **Button** class also includes a **wait_for_press** method which pauses the script until the button is pressed. Open a new file in Python 3 IDLE, enter the code from **ch3listing2.py** (page 23), and save it. This will only print the message at the bottom on the screen once the button has been pressed. The program is then ended.

>STEP-04
Light an LED

Add a red LED to your breadboard, using jumper wires to connect its longer leg to the GPIO 25 pin, and its shorter leg via a resistor to the ground rail; your circuit should resemble the diagram on page 19. In a new Python 3 IDLE file, enter the code from **ch3listing3.py** and save it. At the top, we import the **LED** and **Button** classes from GPIO Zero, along with the **pause** function from **signal**. We then allocate variables to the

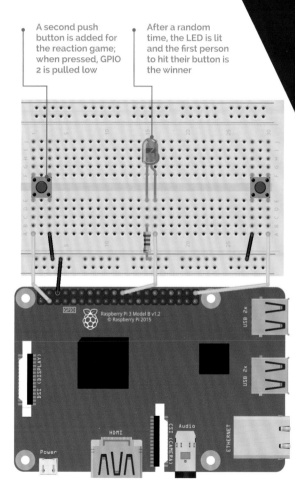

A second push button is added for the reaction game; when pressed, GPIO 2 is pulled low

After a random time, the LED is lit and the first person to hit their button is the winner

Right
The LED is lit! In this reaction game, the first person to now press their button will win

LED and button on GPIO pins 25 and 21 respectively. When the button is pressed, the LED is turned on; when released, it's turned off.

It's also possible to keep the LED lit for a set period after pressing. Open a new file, enter the code from **ch3listing4.py** (page 24), and save it. This time, we wait for a button press as in step 3, then turn the LED on for three seconds, then off.

>STEP-05
Reaction game

By adding a second push button to our circuit, we can make a simple two-player reaction game. When the LED turns on at a random time, the first person to hit their button is the winner. Position the extra button on the breadboard as in the diagram on page 21, connecting it to the ground rail and GPIO 2; move the LED and its connections to the middle, if not there already. Open a new file in Python 3 IDLE, enter the code from **ch3listing5.py** (page 24), and save it. At the top, we import the classes required as before, along with the **random** module. We assign variables to the LED and two buttons, then create a **time** variable equal to a random number between 5 and 10; after sleeping for this number of seconds, the LED is turned on. The **while True:** loop is terminated with **break** when someone presses their button, after printing the appropriate victory message.

ch3listing1.py

Language
>PYTHON 3

DOWNLOAD:
magpi.cc/2bhwcLz

```
from gpiozero import Button

button = Button(21)

while True:
    if button.is_pressed:
        print("Button is pressed")
    else:
        print("Button is not pressed")
```

ch3listing2.py

```
from gpiozero import Button

button = Button(21)

button.wait_for_press()
print("Button was pressed")
```

ch3listing3.py

```
from gpiozero import LED, Button
from signal import pause

led = LED(17)
button = Button(21)

button.when_pressed = led.on
button.when_released = led.off

pause()
```

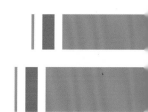

ch3listing4.py

```python
from gpiozero import LED, Button
from time import sleep

led = LED(25)
button = Button(21)

button.wait_for_press()
led.on()
sleep(3)
led.off()
```

ch3listing5.py

```python
from gpiozero import Button, LED
from time import sleep
import random

led = LED(25)

player_1 = Button(21)
player_2 = Button(2)

time = random.uniform(5, 10)
sleep(time)
led.on()

while True:
    if player_1.is_pressed:
        print("Player 1 wins!")
        break
    if player_2.is_pressed:
        print("Player 2 wins!")
        break

led.off()
```

The
MagPi
ESSENTIALS

[**CHAPTER** FOUR]
MAKE A PUSH BUTTON
MUSIC BOX

Use two or more tactile push buttons
to play different sound samples

So far, we've added a push button to a simple circuit to light an LED, and then added a second button to make a reaction game. In this chapter, we'll use several push buttons to make a GPIO music box that triggers different sounds when we press different buttons. For this, we'll make use of GPIO Zero's Button class again, as well as using the Python dictionary structure to assign sounds to buttons.

>STEP-01
Get some sounds

Before we start building our GPIO music box circuit, we'll need to prepare some sound samples for it to play. First, open a terminal window and create a new folder called musicbox for this project: **mkdir musicbox**. Then change to that directory: **cd musicbox**. Now we need to source some sound samples. While there are many public domain sounds to be found online, for this example we'll use some of Scratch's

Right
Each time you press a button, the assigned sound sample will play through a connected speaker

built-in percussion sounds, already present on the Pi. In your terminal, enter **mkdir samples**, then change to that directory: **cd samples**. Now copy the Scratch percussion sounds with:

```
cp /usr/share/scratch/Media/Sounds/Percussion/* .
```

>STEP-02
Play a drum

We'll now create a simple Python program to play a drum sample repeatedly, to check everything is working. Open IDLE from the Main Menu: Menu > Programming > Python 3 (IDLE). Create a new file by clicking File > New File. Now enter the code from the listing **ch4listing1.py** (page 30), changing the WAV file name to suit your own sample if you're using different ones. Save the file in your **musicbox** folder with File > Save.

At the start of the program, we import the **mixer** module from the Pygame library, then its **Sound** class which enables multichannel sound playback in Python. Next, we add a line to initialise the Pygame mixer: **pygame.mixer.init()**. We then create a **Sound** object for one of the files in our samples folder: **drum = Sound("samples/ DrumBuzz.wav")**.

Finally, we add a **while True:** loop to repeatedly play the drum sound. Press **F5** to run

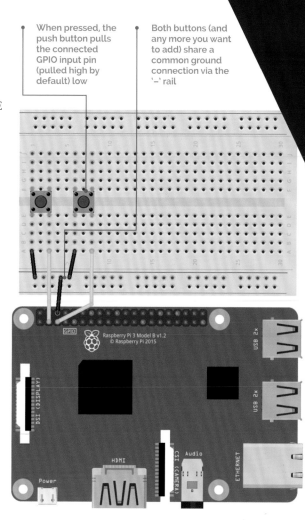

When pressed, the push button pulls the connected GPIO input pin (pulled high by default) low

Both buttons (and any more you want to add) share a common ground connection via the '−' rail

Above Extra buttons can easily be added to the circuit to play more sounds assigned in the Python code

the program and listen to it play. If you can't hear it, you might need to alter your audio configuration; in a terminal, enter `amixer cset numid=3 1` to switch it to the headphone socket, or `amixer cset numid=3 1` to switch to the HDMI output.

>STEP-03
Wire up a button

As usual, it's best to turn the Raspberry Pi off while connecting our circuit on the breadboard. First, we'll add a single button. As before, place the button so it straddles the central groove of the breadboard. One leg is connected to GPIO pin 2, and the other to the common ground rail on the breadboard, which in turn is wired to a GND pin.

We'll now make a sound play whenever the button is pressed. Open a new file in Python 3 IDLE, enter the code from **ch4listing2.py** (page 31), and save it in your **musicbox** folder. At the start of the program, we also import the **Button** class from GPIO Zero, and the **pause** class from the **signal** library. We assign the button variable to GPIO pin 2, with **button = Button (2)**. We then tell the sound to play when the button is pressed:

```
button.when_pressed = drum.play
```

Finally, we add **pause()** at the end so that the program will continue to wait for the button to be pressed. Run the program and every time you press the button, the drum sound should play.

>STEP-04
Add a second button
We'll add a second button to the circuit, so it should now look like the diagram on page 27. Place it on the breadboard as before, and wire it up to GPIO 3 and the common ground rail. Now open a new file in Python 3 IDLE, enter the code from **ch4listing3.py** (page 31), and save it in your **musicbox** folder. Note that rather than assigning the **Button** objects and sounds to the pins individually, we're using a dictionary structure to assign their numbers to sound samples:

```
sound_pins = {
    2: Sound("samples/DrumBizz.wav"),
    3: Sound("samples/CymbalCrash.wav"),
}
```

We then create a list of buttons on all the pin numbers in the **sound_pins** dictionary:

```
buttons = [Button(pin) for pin in sound_pins]
```

Finally, we create a **for** loop that looks up each button in the dictionary and plays the appropriate sound:

```
for button in buttons:
    sound = sound_pins[button.pin.number]
    button.when_pressed = sound.play
```

Run the program and press each button to hear a different sound.

>STEP-05
Add more buttons

The way we have structured the program makes it easy to add extra buttons and assign them to sound samples. Just connect each button to a GPIO number pin (not any other type) and the ground rail, as before. Then add the GPIO pin numbers and sounds to the dictionary, as in the following example:

```python
sound_pins = {
    2: Sound("samples/DrumBizz.wav"),
    3: Sound("samples/CymbalCrash.wav"),
    4: Sound("samples/Gong.wav"),
    14: Sound("samples/HandClap.wav"),
}
```

ch4listing1.py

```python
import pygame.mixer
from pygame.mixer import Sound

pygame.mixer.init()

drum = Sound("samples/DrumBuzz.wav")

while True:
    drum.play()
```

Language
>PYTHON 3

DOWNLOAD:
magpi.cc/zbhwqlH

ch4listing2.py

```python
from gpiozero import Button
import pygame.mixer
from pygame.mixer import Sound
from signal import pause

pygame.mixer.init()
button = Button(2)
drum = Sound("samples/DrumBuzz.wav")

button.when_pressed = drum.play
pause()
```

ch4listing3.py

```python
from gpiozero import Button
import pygame.mixer
from pygame.mixer import Sound
from signal import pause

pygame.mixer.init()

sound_pins = {
    2: Sound("samples/DrumBuzz.wav"),
    3: Sound("samples/CymbalCrash.wav"),
}

buttons = [Button(pin) for pin in sound_pins]
for button in buttons:
    sound = sound_pins[button.pin.number]
    button.when_pressed = sound.play

pause()
```

[CHAPTER FIVE]
MEASURE
CPU USAGE
WITH AN RGB LED

Learn how to use an RGB LED and get it to show CPU load

You'll Need

> GPIO Zero

> 1× solderless breadboard

> 1× RGB LED

> 3× 100Ω resistor

> 4× male-to-female jumper wires

We lit up a standard LED in chapter 2, using GPIO Zero's **LED** class. It also features a special **RGBLED** class for controlling - guess what - an RGB LED! In this chapter, we'll make use of this to light up our LED in different shades by altering the red, green, and blue values. Then we'll code up a little program that tracks the Pi's CPU usage percentage, and adjust the LED between green and red accordingly to show how much processing power it's using.

>STEP-01
Select your RGB LED

Light-emitting diodes (LEDs) are cool. Literally. Unlike a normal incandescent bulb, which has a hot filament, LEDs produce light solely by the movement of electrons in a semiconductor material. An RGB LED has three single-colour LEDs combined in one package. By varying the brightness of each component, you can produce a range of colours, just like mixing paint. There are two main types of RGB LEDs: common anode and common cathode. We're going to use common cathode for this project.

Common cathode RGB LED. The longest leg is the cathode – connect it to ground

The resistors limit the current flowing through the LED and prevent damage to the Raspberry Pi

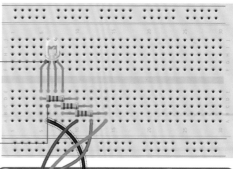

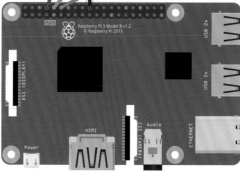

>STEP-02
Connect the RGB LED

As usual, it's best to turn the Raspberry Pi off while connecting our circuit on the breadboard. LEDs need to be connected the correct way round. For a common cathode RGB LED, you have a single ground wire – the longest leg – and three anodes, one for each colour. To drive these from a Raspberry Pi, connect each anode to a GPIO pin via a current-limiting resistor. When one or more of these pins is set to HIGH (3.3V), the LED will light up the corresponding colour. Connect everything as shown in the diagram on page 33.

Here, we wire the cathode (long leg) to a GND pin, while the other legs are wired via resistors to GPIO 14, 15, and 18. The resistors are essential to limit the amount of current flowing to the Pi, to avoid damaging it; we've used 100Ω resistors, but you could get away with using ones with a slightly higher ohmage, such as 330Ω.

Below By altering the three RGB values, you can light the LED in any shade you like

Left Here we're monitoring the Pi's CPU usage; yellow means it's at about 50%

>STEP-03
Test the LED

With the **RGBLED** class in GPIO Zero, it's easy to alter the colour of the LED by assigning values of between 0 and 1 to red, green, and blue. On the Pi, open IDLE from the Main Menu: Menu > Programming > Python 3 (IDLE). Create a new file by clicking File > New file, then enter the code from **ch5listing1.py** (on page 37) and save it.

At the top, we import the **RGBLED** class from GPIO Zero, along with the **sleep** function from the **time** library. We then assign the variable **led** to the **RGBLED** class on GPIO pins 14, 15, and 18, for red, green, and blue. We then make **led.red** equal to 1 to turn the LED a full red colour. After a second, we then change the value to 0.5 to reduce its intensity. We then go through a sequence of colours using **led.color**, assigning it a tuple of red, green, and blue values to mix the shades. So, **(1, 0, 1)** shows full red and blue to make magenta. You can vary each value between 0 and 1 to create an almost infinite range of shades. Finally, we use a **for** loop to slowly increase the intensity of blue.

>STEP-04
Add a new library

We now want to get our RGB LED to change colour between green and red, to show the CPU usage of the Raspberry Pi to which it's connected, so we can track how much of its processing power we're using at any time. For this, we'll need the **psutil** library, which can be installed from the terminal with:

```
sudo pip3 install psutil --upgrade
```

This will let us look up the CPU usage of the Raspberry Pi as a percentage number, which can then be used in our code to vary the LED's colour.

>STEP-05
Measure CPU usage

In IDLE, create a new file, enter the code from **ch5listing2.py**, and save it. At the top, we import the modules we need, including **psutil**. We then assign the **myled** variable to the **RGBLED** class on GPIO 14, 15, and 18, for red, green, and blue. In a never-ending **while True:** loop, we assign the **cpu** variable to the percentage of CPU usage via **psutil**, then assign the red and green LED values accordingly, and light the LED.

Try running the code. The LED should light up: its colour will indicate how hard your Pi's CPU is working. Green means less busy, turning redder as the CPU becomes more heavily loaded. Start up some other applications to test it. If you have an original Model B, you'll probably find that just running *Minecraft* is enough to turn the LED red. If you have a Pi 3, you may need to start lots of things running in order to have any impact!

>STEP-06
Customise your project

The example code only uses the red and green components of the LED: the blue value is always set to zero. You could swap things around and create a different colour gradient (e.g. blue to red), or put together a fancy function that maps a percentage value onto all three colours. Have fun with the colours and maybe even have it look at other resources to monitor...

Language
>PYTHON 3

DOWNLOAD:
magpi.cc/2bhwsdc

ch5listing1.py

```python
from gpiozero import RGBLED
from time import sleep

led = RGBLED(14,15,18)

led.red = 1  # full red
sleep(1)
led.red = 0.5  # half red
sleep(1)

led.color = (0, 1, 0)  # full green
sleep(1)
led.color = (1, 0, 1)  # magenta
sleep(1)
led.color = (1, 1, 0)  # yellow
sleep(1)
led.color = (0, 1, 1)  # cyan
sleep(1)
led.color = (1, 1, 1)  # white
sleep(1)

led.color = (0, 0, 0)  # off
sleep(1)

# slowly increase intensity of blue
for n in range(100):
    led.blue = n/100
    sleep(0.1)
```

ch5listing2.py

```python
from gpiozero import RGBLED
import psutil, time

myled = RGBLED(14,15,18)

while True:
    cpu = psutil.cpu_percent()
    r = cpu / 100.0
    g = (100 - cpu)/100.0
    b = 0
    myled.color = (r, g, b)
    time.sleep(0.1)
```

[CHAPTER SIX]

MAKE A
MOTION-
SENSING
ALARM

Stop people from sneaking up on your stuff by
creating a motion-sensing alarm that buzzes
when it detects someone

You'll Need

> 1× HC-SR501 PIR sensor
> magpi.cc/
> 1rwsEL7

> 1× Mini piezo buzzer
> magpi.cc/
> 1rwsXG2

> Jumper wires

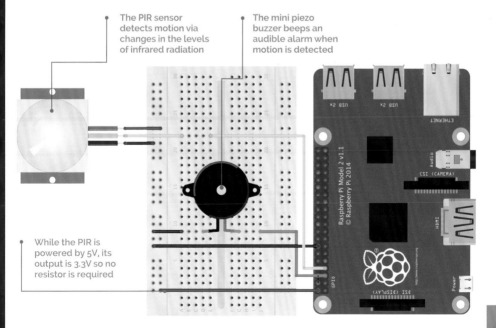

Need to protect your room or precious items from miscreants or nosy family members? With just a PIR motion sensor and a buzzer wired up to your Raspberry Pi, it's very simple to create an intruder alert. Whenever movement is detected in the area, a loud beeping noise will raise the alarm. To take things further, you could add a flashing LED, an external speaker to play a message, or even a hidden Camera Module to record footage of intruders.

>STEP-01
Attach PIR motion sensor

First, we need to wire the PIR (passive infrared) sensor to the Pi. While it could be hooked to the GPIO pins directly using female-to-female jumper wires, we're doing it via a breadboard. The sensor has three pins: VCC (voltage supply), OUT (output), and GND (ground). Use female–to–male jumpers to connect VCC to the '+' rail of the breadboard, and GND to the '−' rail. Wire OUT to a numbered row, then use another jumper to connect that row to GPIO pin 4.

The PIR sensor detects motion via changes in the levels of infrared radiation

The mini piezo buzzer beeps an audible alarm when motion is detected

While the PIR is powered by 5V, its output is 3.3V so no resistor is required

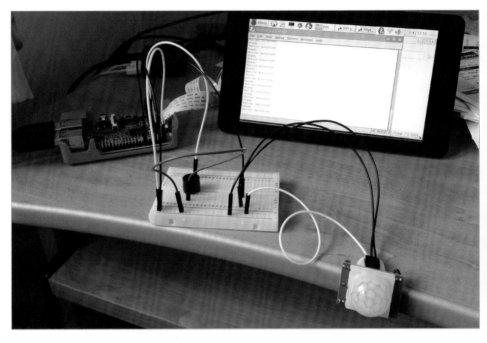

>STEP-02
Wire up the buzzer

Next, we'll hook up the mini buzzer. Place its two legs across the central groove in the breadboard. Note that the longer leg is the positive pin; wire its numbered row to GPIO pin 3 on the Pi to connect it. Wire the row of the buzzer's shorter leg to the '−' rail, then connect the latter to a GND pin on the Pi. Finally, connect the '+' rail to the Pi's 5V pin to power the PIR sensor.

>STEP-03
Work on the code

In IDLE, enter the code from **ch6listing1.py**. At the start, we import the **MotionSensor** and **Buzzer** modules from GPIO Zero, each of which contains numerous useful functions; we'll need a few of them for our intruder alarm. We also import the **time** library so that we can add a delay to the detection loop. Next, we assign the relevant GPIO pins for the PIR sensor and buzzer; we've used GPIO 4 and 3 respectively in this example, but you could use alternatives if you prefer.

Language
>PYTHON 3

DOWNLOAD:
magpi.cc/2bhwrWQ

>STEP-04
Setting things up

Before starting our motion detection **while** loop, we make use of the GPIO Zero library's **wait_for_no_motion** function to wait for the PIR to sense no motion. This gives you time to leave the area, so that it doesn't immediately sense your presence and raise the alarm when you run the code! Once the PIR has sensed no motion in its field of view, it will print 'Ready' on the screen and the motion detection loop can then commence.

ch6listing1.py

```python
from gpiozero import
MotionSensor, Buzzer
import time

pir = MotionSensor(4)
bz = Buzzer(3)

print("Waiting for PIR to settle")
pir.wait_for_no_motion()

while True:
    print("Ready")
    pir.wait_for_motion()
    print("Motion detected!")
    bz.beep(0.5, 0.25, 8)
    time.sleep(3)
```

>STEP-05
Motion detection loop

Using **while True:** means this is an infinite loop that will run continually, until you stop the program by clicking the 'X' icon of its window or pressing **CTRL+C** on the keyboard. Whenever motion is detected by the PIR sensor, we get the buzzer to beep repeatedly eight times: 0.5 seconds on, 0.25 seconds off, but you can alter the timings. We then use **time.sleep(3)** to wait 3 seconds before restarting the loop.

>STEP-06
Adjust the sensitivity

If you find that the alarm is going off too easily or not at all, you may need to adjust the sensitivity of the PIR sensor. This is achieved by using a small screwdriver to adjust the plastic screw of the left potentiometer, labelled Sx; turn it anticlockwise to increase sensitivity. The other potentiometer, Tx, alters the length of time the signal is sent after detection; we found it best to turn it fully anticlockwise, for the shortest delay of 1 second.

[CHAPTER SEVEN]

MAKE A
RANGE
FINDER

Link together an ultrasonic distance
sensor and seven-segment display
to measure distances

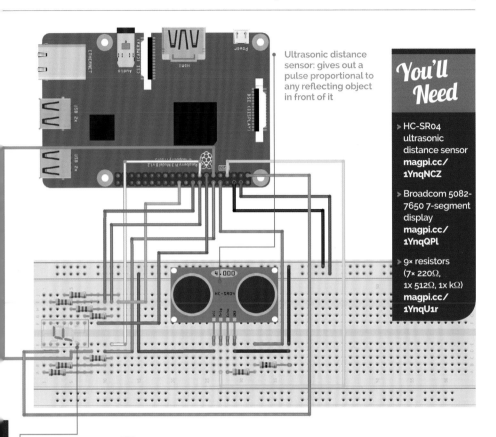

Ultrasonic distance sensor: gives out a pulse proportional to any reflecting object in front of it

You'll Need

> HC-SR04 ultrasonic distance sensor
magpi.cc/ 1YnqNCZ

> Broadcom 5082-7650 7-segment display
magpi.cc/ 1YnqQPl

> 9× resistors (7× 220Ω, 1x 512Ω, 1x kΩ)
magpi.cc/ 1YnqU1r

You can use any seven-segment display, but alternatives might have a different pinout

T he HC-SR04 ultrasonic distance sensor is a great favourite with Pi robot-makers. It works by bouncing ultrasonic sound off an object and timing how long it takes for the echo to return. This time is then converted into a distance which can be displayed on a single-digit, seven-segment display. Here you can acquire the skills to handle inputs and outputs. You also get to use seven-segment displays, which are quite cool in a retro kind of way.

>STEP-01
Lighting the display

The seven-segment display is a collection of LEDs, with one LED corresponding to one of the segments. All the anodes (positive ends) are connected together; this should be connected to the 3V3 supply. The cathodes (negative ends) should be connected to a resistor to limit

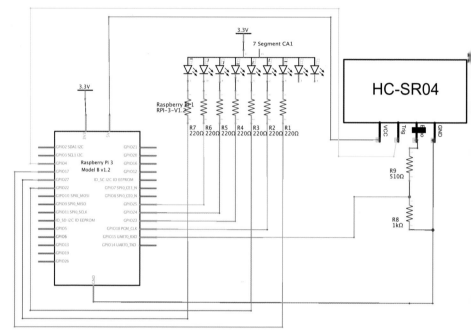

the LED current, and the other end of the resistor to a GPIO pin. To turn the LED on, all you have to do is set the GPIO output to be LOW (0V) and it will complete the circuit for the current to flow.

>STEP-02
Generating a seven-segment pattern

The display consists of four bars or segments that can be lit. By choosing the segments to light up, you can display a number from 0 to 15, although you have to resort to letters (also known as hexadecimal numbers) for this. There are, in fact, 128 different patterns you can make, but most are meaningless. In our code (**ch7listing1.py** overleaf), a list called **seg** defines what pins are connected to what segments, and another list called **segmentPattern** defines the LED pattern for each number.

>STEP-03
Displaying numbers

The **display** function sets up the segments to display any single-digit number passed to it. First, it sets all the segments to off, and then if the number is less than 16, it goes through the entries in the **segmentPattern** list for that number and turns on the appropriate segments. Note that we can still use on and off even though they're not powered by individual GPIO pins, because the LEDs were declared to GPIO Zero as `active_high = False`.

>STEP-04
The distance sensor

The HC-SR04 distance sensor signals its reading by producing an output pulse that the Pi tries to measure. The GPIO Zero library measures this pulse and converts it into a distance by returning a floating-point number that maxes out at 1 metre. We then multiply this number by 10 to give decimetres. Next, we convert it to an integer to get rid of the fractional part of the measurement, so we can show it on our single-digit display.

>STEP-05
Building the project

For our build, we used a dinky little breadboard shield from Dtronixs. This allowed for a much more compact arrangement than a conventional breadboard, although you can of course still use one. As the HC-SR04 uses a 5V power supply, the pulse we have to measure is also nominally 5V. Therefore, this has to be cut down to 3V3 by using a 512Ω and 1kΩ resistor voltage divider.

Below The project in action; the Pi is measuring how far it is to the Raspberry Pi 3 box

>STEP-06
Using the sensor

The distance to the reflective object is updated every 0.8 seconds. If this is greater than a metre, then the display will be blank. A display of 0 indicates that the object is less than 10cm away. Don't touch the sensor, otherwise its readings will be wrong. Also, as it has quite a wide beam, you can get reflections from the side. If several objects are in the field of view, then the distance to the closest one is returned.

ch7listing1.py

Language
>PYTHON 3

DOWNLOAD:
magpi.cc/2aNx9yf

```python
# displays the distance in decimetres on a 7-segment display
from gpiozero import LED
from gpiozero import DistanceSensor
import time

seg = [LED(27,active_high=False),LED(25,active_high=False),LED(24,active_high=False),
       LED(23,active_high=False),LED(22,active_high=False),LED(18,active_high=False),
       LED(17,active_high=False)]

segmentPattern = [[0,1,2,3,4,5],[1,2],[0,1,6,4,3],[0,1,2,3,6],[1,2,5,6],[0,2,3,5,6], #0 to 5
                  [0,2,3,4,5,6],[0,1,2],[0,1,2,3,4,5,6],[0,1,2,5,6],[0,1,2,4,5,6], #6 to A
                  [2,3,4,5,6],[0,3,4,5],[1,2,3,4,6],[0,3,4,5,6],[0,4,5,6] ] #B to F

sensor = DistanceSensor(15,4)

def main() :
    print("Display distance on a 7-seg display")

    while 1:
        distance = sensor.distance * 10 # distance in decimeters
        print("distance",distance)
        if distance >= 10.0:
            distance = 16.0
        display(int(distance))
        time.sleep(0.8)

def display(number):
    for i in range(0,7):
        seg[i].off()
    if number < 16:
        for i in range(0,len(segmentPattern[number])):
            seg[segmentPattern[number][i]].on()

# Main program logic:
if __name__ == '__main__':
    main()
```

[CHAPTER EIGHT]
MAKE A LASER
TRIPWIRE

Learn how to use an LDR to detect
a laser pointer beam

You'll Need

> GPIO Zero

> 1× solderless breadboard

> 1× light-dependent resistor (LDR)

> 1× 1µF capacitor

> 1× laser pointer

> 5× male-to-female jumper wires

> 5× female-to-female jumper wires (optional)

> 1× drinking straw

> 1× plastic box

T he Raspberry Pi can easily detect a digital input via its GPIO pins: any input that's approximately below 1.8V is considered off, and anything above 1.8V is considered on. An analogue input can have a range of voltages from 0V up to 3.3V, however, and the Raspberry Pi is unable to detect exactly what that voltage is. One way of getting around this is by using a capacitor and timing how long it takes to charge up above 1.8V. By placing a capacitor in series with a light-dependent resistor (LDR), the capacitor will charge at different speeds depending on whether it's light or dark. We can use this to create a laser tripwire!

>STEP-01
Connect the LDR

An LDR (also known as a photocell) is a special type of electrical resistor whose resistance is very high when it's dark, but reduced when light is shining on it. With the Raspberry Pi turned off, place your LDR into the breadboard, then add the capacitor. It's essential to

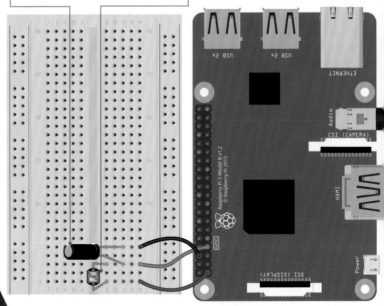

The LDR is connected to a capacitor: the time taken to charge this can be measured

The light-dependent resistor (LDR) has lower resistance when light is shining on its head

get the correct polarity for the latter component: its longer (positive) leg should be in the same breadboard column as one leg of the LDR. Now connect this column (with a leg of both components) to GPIO 4. Connect the other leg of the LDR to a 3V3 pin, and the other leg of the capacitor to a GND pin. Your circuit should now resemble the diagram on page 48.

Above
Shining the laser onto the LDR in a darkened room will dramatically affect the measured light level

>STEP-02
Test the LDR

On the Pi, open IDLE from the Main Menu: Menu > Programming > Python 3 (IDLE). Create a new file by clicking File > New File, enter the code from **ch8listing1.py** (page 52), then save it. At the start, we import the **LightSensor** class from GPIO Zero. We then assign the variable **ldr** to the LDR input on the GPIO 4 pin. Finally, we use a never-ending **while True:** loop to continually display the current value of the light sensed by the LDR, which ranges from 0 to 1. Try running the code and then shining your laser pointer on it to vary the light level.

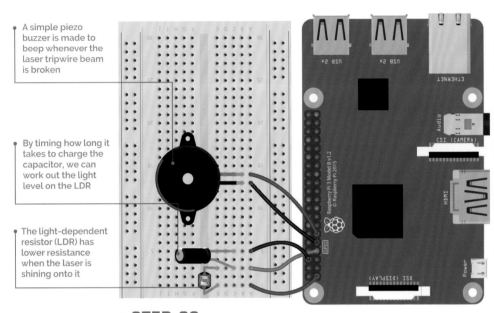

A simple piezo buzzer is made to beep whenever the laser tripwire beam is broken

By timing how long it takes to charge the capacitor, we can work out the light level on the LDR

The light-dependent resistor (LDR) has lower resistance when the laser is shining onto it

>STEP-03
Enclose the LDR

Unless you're working in a darkened room, you'll probably notice little difference between the measured light level when the laser pointer is directed onto the LDR and when it's not. This can be fixed by reducing the amount of light that the LDR receives from other light sources in the room, which will be essential for our laser tripwire device to work effectively. We'll achieve this by cutting off a short section – between 2cm and 5cm – of an opaque drinking straw, and inserting the head of the LDR into one end. Now try the test code again and see how the measured light level changes when you shine the laser pointer into the other end of the straw. You should notice a larger difference in values.

>STEP-04
Wire up the buzzer

To create an audible alarm for our laser tripwire, we'll add a piezo buzzer to the circuit. Again, the polarity has to be correct: connect the column of the buzzer's longer leg to GPIO 17, and the shorter leg to a GND pin. Let's test it's working. In IDLE, create a new file, enter the code from **ch8listing2.py** (page 52), and save it. At the top, we import the **Buzzer** class from GPIO Zero. Next, we assign the **buzzer** variable to the buzzer output on GPIO 17. Finally, we use **buzzer.beep** to make the buzzer turn on and off repeatedly at the default length of 1 second. To stop it, close the Python shell window when it's off.

>STEP-05
Test the tripwire

We'll now put everything together so that the laser pointer shines at the LDR (through the straw) and whenever the beam is broken, the buzzer sounds the alarm. In IDLE, create a new file, enter the code from **ch8listing 3.py** (page 52), and save it. At the start, we import the **Buzzer** and **LightSensor** classes from GPIO Zero. We also import the **sleep** function from **time**; we'll need this to slow the script down a little to give the capacitor time to charge. As before, we assign variables for the buzzer and LDR to the respective devices on GPIO pins 4 and 17. We then use a **while True:** loop to continually check the light level on the LDR; if it falls below 0.5, we make the buzzer beep. You can change this number to adjust the sensitivity; a higher value will make it more sensitive. Try running the code; if you break the laser beam, the buzzer should beep for 8 seconds. You can adjust this by altering the **buzzer.beep** parameters and **sleep** time.

>STEP-06
Package it up

Once everything is working well, you can enclose your Raspberry Pi and breadboard in a plastic box (such as an old ice cream tub), with the drinking straw poking through a hole in the side. If you prefer, you can remove the breadboard and instead connect the circuit up directly by poking the legs of the components into female-to-female jumper wires, with the long capacitor leg and an LDR leg together in one wire end, connected to the relevant pins. Either way, place the tub near a doorway with the laser pointer on the other side, with its beam shining into the straw. Run your code and try walking through the doorway: the alarm should go off!

Below
Place your laser tripwire across a doorway; when someone breaks the beam, the alarm will sound

Language

>PYTHON 3

DOWNLOAD:
magpi.cc/2bhxwxC

ch8listing1.py

```python
from gpiozero import LightSensor

ldr = LightSensor(4)

while True:
    print(ldr.value)
```

ch8listing2.py

```python
from gpiozero import Buzzer

buzzer = Buzzer(17)
buzzer.beep()
```

ch8listing3.py

```python
from gpiozero import LightSensor, Buzzer
from time import sleep

ldr = LightSensor(4)
buzzer = Buzzer(17)

while True:
    sleep(0.1)
    if ldr.value < 0.5:
        buzzer.beep(0.5, 0.5, 8)
        sleep(8)
    else:
        buzzer.off()
```

The
MagPi
ESSENTIALS

[CHAPTER NINE]

BUILD AN
INTERNET
RADIO

Use potentiometers to control an
LED and tune in to radio stations

You'll Need

> GPIO Zero

> 1× solderless breadboard

> 1× MCP3008 ADC chip

> 2× potentiometers

> 1× LED

> 1× 330Ω resistor

> 7× male-to-female jumper wires

> 10× male-to-male jumper wires

Another way for the Raspberry Pi to detect analogue inputs is by using an analogue-to-digital converter (ADC) chip, such as the MCP3008. The latter offers eight input channels to connect sensors and other analogue inputs. In this tutorial, we'll hook up a potentiometer to an MCP3008, to control the brightness of an LED by turning the knob. We'll then add a second potentiometer and create an internet radio, using the two potentiometers to switch the station and alter the volume.

>STEP-01
Enable SPI
The analogue values from the ADC chip will be communicated to the Pi using the SPI protocol. While this will work in GPIO Zero out of the box, you may get better results if you enable full SPI support. To do this, open a terminal window and enter:

```
sudo apt-get install python3-spidev python-spidev
```

Click OK and reboot the Pi.

- Four legs are connected to special GPIO pins, while the rest are hooked up to the power and ground rails

- The MCP3008 ADC chip straddles the central groove; the side shown without wires comprises eight input channels

Left
By turning the potentiometer knob, we can adjust the brightness of the LED

>STEP-02
Connect the ADC

As usual, it's best practice to turn off the Pi while creating our circuit. As you can see from the diagram, there's quite a lot of wiring required to connect the MCP3008 ADC to the Pi's GPIO pins. As an alternative, you could use an Analog Zero (**magpi.cc/2aey1b6**), which provides the MCP3008 chip on a handy add-on board.

First, place the MCP3008 in the middle of the breadboard, straddling its central groove. Now connect the jumper wires as in the diagram. Two go to the '+' power rail, connected to a 3V3 pin; two others are connected to a GND pin via the '−' rail. The four middle legs of the ADC are connected to GPIO pins 8 (CE0), 10 (MOSI), 9 (MISO), and 11 (SCLK).

>STEP-03
Read the value

Now the ADC is connected to the Pi, you can wire devices up to its eight input channels (numbered 0 to 7). Here, we'll connect a potentiometer, which is a variable resistor: as you turn its rotary knob, the Pi reads the voltage (from 0V to 3.3V). We can use this for precision control of other

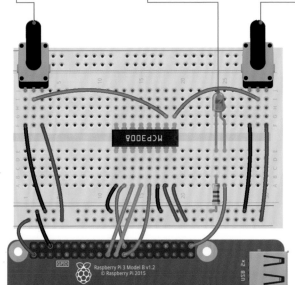

Connected to channel 1 of the MCP3008, the second potentiometer adjusts the volume of our radio

The LED is connected as normal, with its long leg wired to GPIO 21 and the other to the ground rail via a resistor

This potentiometer is connected to channel 0 of the MCP3008; on our radio, it's used to switch the station

components, such as an LED. As in the diagram, connect one outer leg of the potentiometer (top-right) to the '−' ground rail, the other side to the '+' power rail, and the middle leg to the first input of the MCP3008: channel 0.

>STEP-04
Read the value

We can now read the potentiometer's value in Python. On the Pi, open IDLE from the Main Menu: Menu > Programming > Python 3 (IDLE). Create a new file by clicking File > New file, then enter the code from **ch9listing1.py** (page 59), and save it. At the top we import the **MCP3008** class from GPIO Zero, then assign the **pot** variable to the ADC's channel 0. A **while True:** loop then continuously displays the potentiometer's value (from 0 to 1) on the screen; try turning it as the code runs, to see the number alter.

>STEP-05
Light an LED

Next, we'll add an LED to the circuit as in the diagram, connecting its longer (positive) leg to GPIO 21, and its shorter leg via a resistor to the '−' ground rail. In IDLE, create a new file, enter the code from **ch9listing2.py** (page 59), and save it. At the start, we import the

MCP3008 and **PWMLED** classes. The latter enables us to control the brightness of an LED using pulse-width modulation (PWM). We create a **PWMLED** object on GPIO 21, assigning it to the **led** variable. We assign our potentiometer to channel 0, as before. Finally, we use GPIO Zero's clever source and values system to pair the potentiometer with the LED, to continuously set the latter's brightness level to the former's value. Run the code and turn the knob to adjust the LED's brightness.

>STEP-06
Add a second pot

Let's add a second potentiometer to our circuit as in the diagram, with its middle leg connected to channel 1 of the MCP3008. We'll now use both potentiometers to control our LED's blink rate. In IDLE, create a new file, enter the code from **ch9listing3.py** (page 59), and save it. Here, we create two separate **pot1** and **pot2** variables, assigned to the ADC's channels 0 and 1 respectively. In a **while True:** loop, we then print the two values on the screen and make the LED blink, with its on and off times affected by our two potentiometers. Run the code and twist both knobs to see how it changes.

Left Some serious spaghetti wiring is required; alternatively, you could use an Analog Zero board to reduce this

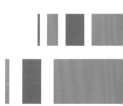

>STEP-07
Install mplayer

We'll use the same circuit to create a simple internet radio, with one potentiometer used to switch the station and the other to adjust the volume. First, we'll need to install the MPlayer media player to be able to play M3U internet radio streams, since Omxplayer can't do this. Open a terminal window and enter:

```
sudo apt-get install mplayer
```

>STEP-08
Make the radio

Open IDLE, create a new file, enter the code from **ch9listing4.py** (page 60), and save it. At the start, we import the **MCP3008** class, along with the **time** and **os** libraries; the latter will enable us to start MPlayer by sending commands directly to the terminal. We create variables for the station and volume dials, on ADC channels 0 and 1 respectively. We then assign variables to two radio stream URLs, for Magic and BBC Radio 1, and assign the **current_station** variable to the latter.

Next, we create a function called **change_station** which includes an **if** condition, so it only triggers when the station set by the first potentiometer position is different from the currently selected one (**current_station**). If so, it stops the current stream and starts playing the new one, before reassigning the **current_station** variable to it.

Finally, in a **while True:** loop, we set the audio volume to the value of the second potentiometer using amixer; we've assigned a minimum value of 65%, but you can alter this. It then checks whether the first potentiometer is below or above 0.5 and calls the **change_station** function.

Run the code and try turning both potentiometers to switch the station and adjust the volume. To keep things simple, we've only used two radio stations in this example, but you could easily add more.

ch9listing1.py

```python
from gpiozero import MCP3008

pot = MCP3008(channel=0)

while True:
    print(pot.value)
```

ch9listing2.py

```python
from gpiozero import MCP3008, PWMLED

pot = MCP3008(0)
led = PWMLED(21)

led.source = pot.values
```

ch9listing3.py

```python
from gpiozero import MCP3008, PWMLED

pot1 = MCP3008(0)
pot2 = MCP3008(1)
led = PWMLED(21)

while True:
    print(pot1.value, pot2.value)
    led.blink(on_time=pot1.value, off_time=pot2.value, n=1, background=False)
```

ch9listing4.py

```python
from gpiozero import MCP3008
import time
import os

station_dial = MCP3008(0)
volume_dial = MCP3008(1)

Magic = "http://tx.whatson.com/icecast.php?i=magic1054.mp3.m3
Radio1 = "http://www.listenlive.eu/bbcradio1.m3u"

current_station = Radio1

def change_station(station):
    global current_station
    if station != current_station:
        os.system("killall mplayer")
        os.system("mplayer -playlist " + station + " &")
        current_station = station

while True:
    vol = (65 + volume_dial.value * 35)
    os.system("amixer set PCM -- " + str(vol) +"%")
    if station_dial.value >= 0.5:
        station = Magic
        change_station(station)
    elif station_dial.value < 0.5:
        station = Radio1
        change_station(station)
    time.sleep(0.1)
```

[CHAPTER TEN]

CREATE AN LED THERMOMETER

Read a temperature sensor and
display its value as a bar graph

Continuing the theme of analogue inputs, we'll use the MCP3008 analogue-to-digital converter (ADC) again and this time hook it up to a temperature sensor. We'll display the current temperature on the screen, then add some LED's and use GPIO Zero's handy **LEDBarGraph** class to get them to light up according to the temperature.

>STEP-01
Enable SPI

As in chapter 9, the analogue values from the ADC chip will be communicated to the Pi using the SPI protocol. While this will work in GPIO Zero out of the box, you may get better results if you enable full SPI support. If you haven't done this already, open a terminal window and enter:

```
sudo apt-get install python3-spidev python-spidev
```

Click OK and reboot the Pi.

Right The TMP36 temperature sensor (bottom-right) is connected to an input channel of the MCP3008 chip

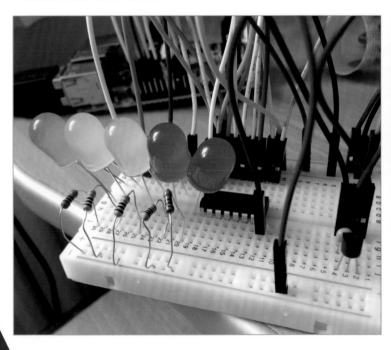

>STEP-02
Connect the ADC

If you already have the MCP3008 wired up from chapter 9, leave it in place, straddling the central groove of the breadboard. As noted before, there's quite a lot of wiring required; as an alternative, you could use an Analog Zero (**magpi.cc/2aey1b6**) add-on board to cut down on this. Otherwise, connect the jumper wires as in the diagram. Two go to the '+' power rail, connected to a 3V3 pin; two others are connected to a GND pin via the '−' rail. The four middle legs of the ADC are connected to GPIO pins 8 (CE0), 10 (MOSI), 9 (MISO), and 11 (SCLK).

>STEP-03
Add the sensor

Now that the ADC is connected to the Pi, you can wire devices up to its eight input channels, numbered 0 to 7. Here, we'll connect a TMP36

It's vital to get the wiring correct for the TMP36 sensor, otherwise it will overheat

Link the sensor's ground and output pins with a capacitor to help stabilise its readings

The MCP3008 ADC chip straddles the central groove; the side shown without wires comprises eight input channels

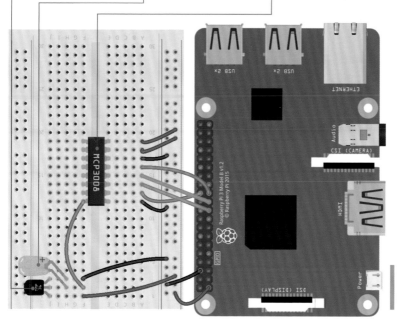

analogue temperature sensor. It's vital that this is wired up correctly, otherwise it'll overheat. With its flat face towards you, the left-hand leg is for power, so connect this to the '+' power rail. The right-hand leg is connected to the '−' ground rail. Its middle leg is the output; here we're connecting to channel 7 (the nearest one) of the MCP3008. Finally, to help stabilise the readings which might otherwise be erratic, we'll add a capacitor to link its output and ground legs.

>STEP-04
Take the temperature

We can now read the sensor's value in Python. On the Pi, open IDLE from the Main Menu: Menu > Programming > Python 3 (IDLE). Create a new file, enter the code from **ch10listing1.py** (page 66), and save it. At the top we import the **MCP3008** class from GPIO Zero, then the **sleep** function from the **time** library. Next, we define a function that

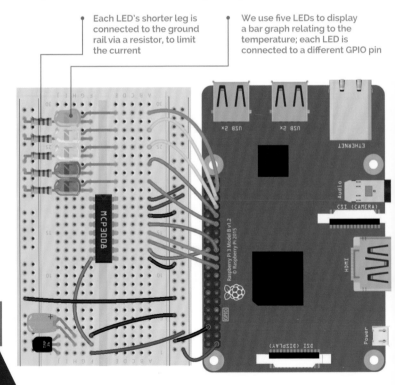

Each LED's shorter leg is connected to the ground rail via a resistor, to limit the current

We use five LEDs to display a bar graph relating to the temperature; each LED is connected to a different GPIO pin

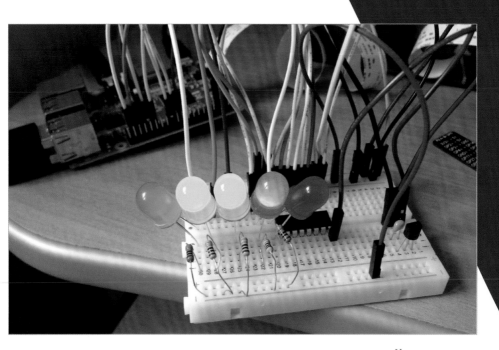

converts the sensor reading into degrees Celsius. We then assign the **adc** variable to channel 7 of the MCP3008. Finally, we use a **for** loop to display the converted temperature on the screen, updating it every second. Note: if you've just been handling the sensor, it might take a little while to settle down.

Above
When we run the final code, the LEDs light up to indicate the temperature read by the sensor

>STEP-05
LED bar graph

Next, we'll add our line of five LEDs to the circuit, as in the diagram. From green to red, we've connected their longer legs to the following GPIO pins: 26, 19, 13, 6, and 5. In IDLE, create a new file, enter the code from **ch10listing2.py** (page 67), and save it. At the start, we import the **LEDBarGraph** class from GPIO Zero; this will enable us to use the LEDs to display a bar graph, saving a lot of complex coding. We assign the **graph** variable to our LEDs on the GPIO pins mentioned previously, and also enable PWM so that we can adjust their brightness for a more accurate display. We then set **graph.value** to various fractions between 0 and 1 to light the relevant number of

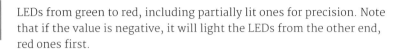

LEDs from green to red, including partially lit ones for precision. Note that if the value is negative, it will light the LEDs from the other end, red ones first.

>STEP-06
Display the temperature

So, we've got our temperature sensor and LED bar graph set up; let's combine them to display the temperature on the LED bar graph. In IDLE, create a new file, enter the code from **ch10listing 3.py**, and save it. At the top, we import GPIO Zero's **MCP3008** and **LEDBarGraph** classes, along with the **sleep** function from the **time** library. As in our original code, we then define a function to convert the temperature sensor's readings to degrees Celsius. We assign the **adc** variable to channel 7 of the MCP3008 and **graph** to our LEDs' GPIO pins, setting PWM to **true**. Finally, in our **for** loop, we add a **bars** variable to determine how many LEDs are lit in the bar graph. In this example, we've divided **temp** by 35, which is around the maximum temperature for the UK, so if it gets to 35°C, all the LEDs should light up fully. Naturally, you can adjust this number to suit your own location's climate. When ready, run the code and see those LEDs light up to show the current temperature.

ch10listing1.py

```python
from gpiozero import MCP3008
from time import sleep

def convert_temp(gen):
    for value in gen:
        yield (value * 3.3 - 0.5) * 100

adc = MCP3008(channel=7)

for temp in convert_temp(adc.values):
    print("The temperature is", temp, "C")
    sleep(1)
```

ch10listing2.py

Language
>PYTHON 3

DOWNLOAD:
magpi.cc/2bhwQbJ

```python
from gpiozero import LEDBarGraph
from time import sleep

graph = LEDBarGraph (26, 19, 13, 6, 5, pwm=True)

graph.value = 1/10
sleep(1)
graph.value = 3/10
sleep(1)
graph.value = -3/10
sleep(1)
graph.value = 9/10
sleep(1)
graph.value = 95/100
sleep(1)
graph.value = 0
```

ch10listing3.py

```python
from gpiozero import MCP3008, LEDBarGraph
from time import sleep

def convert_temp(gen):
    for value in gen:
        yield (value * 3.3 - 0.5) * 100

adc = MCP3008(channel=7)
graph = LEDBarGraph (26, 19, 13, 6, 5, pwm=True)

for temp in convert_temp(adc.values):
    bars = temp / 35
    graph.value = bars
    sleep(1)
```

[CHAPTER ELEVEN]

BUILD A
GPIO ZERO
ROBOT

Control DC motors with GPIO Zero and build a
Pi Zero robot

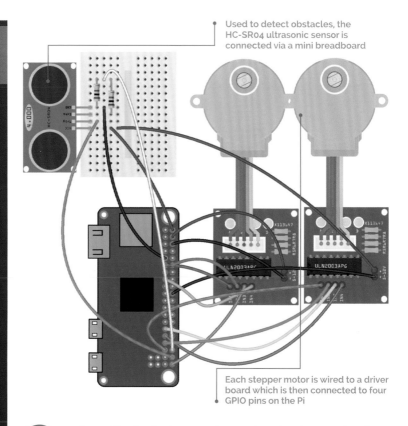

Used to detect obstacles, the HC-SR04 ultrasonic sensor is connected via a mini breadboard

Each stepper motor is wired to a driver board which is then connected to four GPIO pins on the Pi

You'll Need

- GPIO Zero

- 1× 3D-printed KOROBOT shell and wheels, or some craft materials **magpi.cc/ 1PCfwMK**

- 1×HC-SR04 ultrasonic sensor **magpi.cc/ 1PCfAMs**

- 2× 28BYJ-48 stepper motors & ULN2003A driver boards **magpi.cc/ 1PCfCE3**

- 1× half-size solderless breadboard

- Mobile power bank

- Various jumper wires

- 2× DC motors and wheels (optional, for steps 1-3)

Raspberry Pi robotics is a popular pastime, but has previously required some complex coding to steer your bots. Fortunately, GPIO Zero makes it much simpler with its **Motor** and **Robot** classes. We have a play around with these at the start of our guide to control a couple of DC motors, before showing you how to build and program a two-wheeled ZeroBot using a Pi Zero and two stepper motors for precision control.

>STEP-01
Connect DC motors

To enable the Raspberry Pi to control DC motors, an H-bridge motor driver board is required; you should never connect motors directly to the Pi, as this is likely to damage the latter. There are numerous

Above GPIO Zero's Robot class makes it very easy to control a two-wheeled robot like this one

motor drivers available; for steps 1–3 of this guide, we're using the one supplied in the popular CamJam EduKit #3 (**magpi.cc/2algEVU**), which fits onto the Pi's GPIO header. Each motor is connected by two wires going to positive and negative terminals on the driver board, which is hooked up to a power supply such as a battery box or a 5V pin on the Pi. Either way, the Raspberry Pi itself is normally powered separately, using a mobile power bank if you want to use your robot untethered.

>STEP-02
Run a motor

GPIO Zero includes a **Motor** class for running bidirectional motors connected via an H-bridge motor driver circuit. On the Raspberry Pi, open IDLE from the Main Menu: Menu > Programming > Python 3 (IDLE). Create a new file by clicking File > New file, then enter the code from **ch11listing1.py** (page 73). At the top, we import the **Motor** class from GPIO Zero, along with **sleep** from **time**. We then assign the **motor** variable to the **Motor** class on the two GPIO pins connected to our motor: in this example, 8 and 7. When we use **motor.forward()**, the motor should run forwards. Within the brackets, we can add a speed between 0 and 1 (the default). Similarly, **motor.backward** will run it backwards, while **motor.stop** will stop the motor if it's still running.

>STEP-03
Move a robot

While you can control your motors individually using the **Motor** class, GPIO Zero also includes the **Robot** class for controlling a two-wheeled robot. Assuming you have such a robot already assembled, open IDLE, create a new file, enter the code from **ch11listing2.py** (page 74), and save it. At the top, we import the **Robot** class from GPIO Zero, along with **sleep** from the **time** library. We then assign the **robot** variable to it, with the relevant GPIO pins for the left and right motors. We can then run various commands to control it, including telling it to spin left or right. In this example, we're using a **for** loop with forward and turn right directions to make it drive around in a square pattern; adjust the **time.sleep** values to determine the square size. Try altering the directions to make different patterns.

Note: To save plugging your robot's Pi into a display each time, you can SSH into it to control it from another computer, or even a tablet or smartphone, connected to the same wireless network.

>STEP-04
Build a ZeroBot

Next, we'll show you how to build a ZeroBot based on a Pi Zero and two stepper motors. The 28BYJ-48 is a cheap but versatile stepper motor that can normally be bought with a ULN2003A driver board for under £4. Stepper motors can be programmed to move in discrete steps, rather than just turned on/off like servos. Using the Pi Zero, you'll be able to control the speed and positioning of the motors very accurately. To cause the motor to rotate, you provide a sequence of 'high' and 'low' levels to each of the four inputs. The direction can then be reversed by reversing the sequence. In the case of the 28BYJ-48, there are four-step and eight-step sequences. The four-step is faster, but the torque is lower. The example code in **ch11listing3.py** (page 74) lets you specify the number of steps through the **seqsize** variable.

Each motor has a connector block at the end of its coloured wires that slots into the white header on the ULN2003A. The GPIO pins controlling that motor connect to the four input pins below the IC, while the 5V power and ground connections go to the bottom two pins on the right.

>STEP-05
Eyes to see

We'll give our ZeroBot some simple 'eyes' that allow it to detect obstacles, courtesy of the HC-SR04 ultrasonic sensor. This has four pins, including ground (GND) and 5V supply (VCC). Using Python, you can tell the Pi to send an input signal to the Trigger Pulse Input (TRIG) by setting a GPIO pin's output to HIGH. This will cause the sensor to send out an ultrasonic pulse to bounce off nearby objects. The sensor detects these reflections, measures the time between the trigger and returned pulse, and then sets a 5V signal on the Echo Pulse Output (ECHO) pin. Python code can measure the time between output and return pulses. Connect the HC-SR04 as shown in the diagram (page 69). Its ECHO output is rated at 5V, which could damage the

Above
The diminutive ZeroBot features a Pi Zero, two stepper motors, and a 3D-printed chassis

Pi. To reduce this to 3V, use two resistors to create a simple voltage divider circuit, as shown in the diagram.

Once you have all your components connected, you can test the code on a bench before building the full robot. Point the 'eyes' away from you and run the code. The red LEDs on the ULN2003As should flash and both motors should start turning. Our example has the bot move in a square. Check that the motors behave accordingly then rerun the code, but this time place your hand a couple of centimetres in front of the HC-SR04 and check that everything stops.

>STEP-06
Give it a body

Now it's time to give the bot a body. If you have access to a 3D printer, you can print the parts for the ZeroBot. This design fits together easily, although you do need to glue the chassis end-caps in place. Alternatively, you could construct a similar design using reasonably thick cardboard for the wheels and part of a plastic bottle as the main tubular chassis. Use more cardboard for the end-caps.

Put your mobile power bank at the bottom of the chassis tube, then attach the motors to the end-caps with screws. Next, place the ULN2003A boards on top of the power bank, and then sit the breadboard with the HC-SR04 'eyes' on top. Finally, slot the Pi Zero in at the back. All nice and cosy, and ready to roll!

ch11listing1.py

```
from gpiozero import Motor
from time import sleep

motor = Motor(forward=8, backward=7)

while True:
    motor.forward()
    sleep(5)
    motor.backward()
    sleep(5)
```

ch11listing2.py

```python
from gpiozero import Robot
from time import sleep

robot = Robot(left=(8, 7), right=(10, 9))

for i in range(4):
    robot.forward()
    sleep(1)
    robot.right()
    sleep(0.2)
```

ch11listing3.py

```python
import time, sys
from gpiozero import DistanceSensor, OutputDevice
from threading import Thread

sensor = DistanceSensor(echo = 16, trigger = 20)

IN1_m1 = OutputDevice(17)
IN2_m1 = OutputDevice(18)
IN3_m1 = OutputDevice(21)
IN4_m1 = OutputDevice(22)
StepPins_m1 = [IN1_m1,IN2_m1,IN3_m1,IN4_m1] # Motor 1 pins
IN4_m2 = OutputDevice(19)
IN3_m2 = OutputDevice(13)
IN2_m2 = OutputDevice(5)
IN1_m2 = OutputDevice(6)
StepPins_m2 = [IN1_m2,IN2_m2,IN3_m2,IN4_m2] # Motor 2 pins
Seq = [[1,0,0,1], # Define step sequence
       [1,0,0,0], # as shown in manufacturer's datasheet
       [1,1,0,0],
       [0,1,0,0],
       [0,1,1,0],
       [0,0,1,0],
       [0,0,1,1],
       [0,0,0,1]]
```

Language
>PYTHON 3

DOWNLOAD:
magpi.cc/2bhwBxf

```python
StepCount = len(Seq)
all_clear = True
running = True

def bump_watch(): # thread to watch for obstacles
    global all_clear
    while running:
        value = sensor.distance
        if value < 0.1: # trigger if obstacle within 10cm
            all_clear = False
        else:
            all_clear = True
def move_bump(direction='F', seqsize=1, numsteps=2052):
    counter = 0 # 2052 steps = 1 revolution for step size of 2
    StepDir = seqsize # Set to 1 or 2 for fwd, -1 or -2 for back
    if direction == 'B':
        StepDir = StepDir * -1
    WaitTime = 10/float(1000) # adjust this to change speed
    StepCounter = 0
    while all_clear and counter < numsteps: # only move if no obstacles
        for pin in range(0, 4):
            Lpin = StepPins_m1[pin]
            Rpin = StepPins_m2[pin]
            if Seq[StepCounter][pin]!=0: # F=fwd, B=back, L=left, R=right
                if direction == 'L' or direction == 'B' or direction == 'F':
                    Lpin.on() # Left wheel only
                if direction == 'R' or direction == 'B' or direction == 'F':
                    Rpin.on() # Right wheel only
            else:
                Lpin.off()
                Rpin.off()
        StepCounter += StepDir
        if (StepCounter>=StepCount): # Repeat sequence
            StepCounter = 0
        if (StepCounter<0):
            StepCounter = StepCount+StepDir
        time.sleep(WaitTime) # pause
        counter+=1
t1 = Thread(target=bump_watch) # run as separate thread
t1.start() # start bump watch thread
for i in range(4): # Draw a right-handed square
    move_bump('F',-2,4104)
    move_bump('R',-2,2052)
running = False
```

[CHAPTER TWELVE]
QUICK
REFERENCE

To help you get started with GPIO Zero, here's a handy reference guide to its many useful classes that make Python coding for physical computing so much simpler

[GPIO PIN NUMBERS]

Note that all GPIO pin references in this guide use Broadcom (BCM) numbering. Refer to **pinout.xyz** for more details.

01. OUTPUT DEVICES

GPIO Zero includes a range of classes that make it easy to control output components such as LEDs, buzzers, and motors…

LED

```
gpiozero.LED(pin, active_high=True, initial_value=False)
```

Use this class to turn an LED on and off. The LED should have its longer leg (anode) connected to a GPIO pin, and the other leg connected via a limiting resistor to GND. The following example will light an LED connected to GPIO 17:

```
from gpiozero import LED

led = LED(17)
led.on()
```

Methods:

`on()`
Turn the device on.

`off()`
Turn the device off.

`blink(on_time=1, off_time=1, n=None, background=True)`
Make the device turn on and off repeatedly.

toggle()
Reverse the state of the device; if on, it'll turn off, and vice versa.

is_lit
Returns True if the device is currently active, and False otherwise.

pin
The GPIO pin that the device is connected to.

PWMLED

gpiozero.PWMLED(pin, active_high=True, initial_value=0, frequency=100**)**

This class is used to light an LED with variable brightness. As before, a resistor should be used to limit the current in the circuit. The following example will light an LED connected to pin 17 at half brightness:

```
from gpiozero import PWMLED

led = PWMLED(17)
led.value = 0.5
```

Methods:

on()
Turn the device on.

off()
Turn the device off.

blink(on_time=1, off_time=1, fade_in_time=0, fade_out_time=0, n=None, background=True**)**
Make the device turn on and off repeatedly.

```
pulse(fade_in_time=1, fade_out_time=1,
n=None, background=True)
```
Make the device fade in and out repeatedly.

```
toggle()
```
Toggle the state of the device. If it's currently off (value is 0.0), this changes it to 'fully' on (value is 1.0). If the device has a duty cycle (value) of 0.1, this will toggle it to 0.9, and so on.

```
is_lit
```
Returns True if the device is currently active, and False otherwise.

```
pin
```
The GPIO pin that the device is connected to.

```
value
```
The duty cycle of the PWM device, from 0.0 (off) to 1.0 (fully on).

RGBLED

```
gpiozero.RGBLED(red, green, blue, active_high=True,
initial_value=(0, 0, 0), pwm=True)
```

As shown in chapter 5, this class is used to light a full-colour LED (composed of red, green, and blue LEDs). Connect its longest leg (cathode) to GND, and the other to GPIO pins via resistors (or use one on the cathode). The following code will make the LED purple:

```
from gpiozero import RGBLED

led = RGBLED(2, 3, 4)
led.color = (1, 0, 1)
```

Methods:

```
on()
```
Turn the device on: equivalent to setting the LED colour to white (1, 1, 1).

off()
Turn the device off: equivalent to setting the LED colour to black
(0, 0, 0).

blink(`on_time=1, off_time=1, fade_in_time=0, fade_out_time=0, on_color=(1, 1, 1), off_color=(0, 0, 0), n=None, background=True`**)**
Make the device turn on and off repeatedly.

pulse(`fade_in_time=1, fade_out_time=1, on_color=(1, 1, 1), off_color=(0, 0, 0), n=None, background=True`**)**
Make the device fade in and out repeatedly.

toggle()
Toggle the state of the device. If it's currently off (value is (0, 0, 0)), this changes it to 'fully' on (value is (1, 1, 1)). If the device has a specific colour, this method inverts it.

color
Represents the color of the LED as an RGB 3-tuple of (red, green, blue), where each value is between 0 and 1 if pwm=True, and only 0 or 1 if not.
For example, purple is (1, 0, 1), yellow is (1, 1, 0), and orange is (1, 0.5, 0).

is_lit
Returns True if the LED is currently active (not black) and False otherwise.

Buzzer

gpiozero.Buzzer(pin, `active_high=True, initial_value=False`**)**

This class is used to control a piezo buzzer. This example will sound a buzzer connected to GPIO pin 3:

```
from gpiozero import Buzzer

bz = Buzzer(3)
bz.on()
```

Methods:

on()
Turn the device on.

off()
Turn the device off.

beep(on_time=1, off_time=1, n=None, background=True)
Make the device turn on and off repeatedly.

toggle()
Reverse the state of the device; if on, it'll turn off, and vice versa.

is_active
Returns True if the device is currently active, and False otherwise.

pin
The GPIO pin that the device is connected to.

Motor

```
gpiozero.Motor(forward, backward, pwm=True)
```

This class will drive a generic motor connected via an H-bridge motor controller. The following example will make a motor connected to GPIO pins 17 and 18 turn 'forwards':

```
from gpiozero import Motor

motor = Motor(17, 18)
motor.forward()
```

Methods:

backward(speed=1**)**
Drive the motor backwards. Speed can be any value between 0 and 1
(**if pwm=True**).

forward(speed=1**)**
Drive the motor forwards. Speed can be any value between 0 and 1
(**if pwm=True**).

stop()
Stop the motor.

02. INPUT DEVICES

The GPIO Zero module includes a range of classes that make it easy to
obtain values from input devices such as buttons and sensors...

Button

```
gpiozero.Button(pin, pull_up=True, bounce_time=None)
```

Use this class with a simple push button or switch. The following
example will print a line of text when a button connected to GPIO
pin 4 is pressed:

```
from gpiozero import Button

button = Button(4)
button.wait_for_press()
print("The button was pressed!")
```

Methods:

wait_for_press(timeout=None**)**
Pause the script until the device is activated, or the timeout (in seconds) is reached.

wait_for_release(timeout=None**)**
Pause the script until the device is deactivated, or the timeout (in seconds) is reached.

when_pressed
The function to run when the device changes state from inactive to active.

when_released
The function to run when the device changes state from active to inactive.

when_held
The function to run when the device has remained active for **hold_time** seconds.

hold_time
The length of time (in seconds) to wait after the device is activated, until executing the **when_held** handler. If **hold_repeat** is True, this is also the length of time between calls to **when_held**.

hold_repeat
If True, **when_held** will be executed repeatedly with **hold_time** seconds between each call.

held_time
The length of time (in seconds) that the device has been held for.

is_held
When True, the device has been active for at least **hold_time** seconds.

> **is_pressed**
> Returns True if the device is currently active, and False otherwise.
>
> **pin**
> The GPIO pin that the device is connected to.
>
> **pull_up**
> If True, the device uses a pull up resistor to set the GPIO pin 'high' by default.

Line Sensor

gpiozero.LineSensor(pin)

This class is used to read a single pin line sensor, like the TCRT5000 found in the CamJam EduKit #3. The following example will print a line of text indicating when the sensor (with its output connected to GPIO pin 4) detects a line, or stops detecting one:

```
from gpiozero import LineSensor
from signal import pause

sensor = LineSensor(4)
sensor.when_line = lambda: print('Line detected')
sensor.when_no_line = lambda: print('No line detected')
pause()
```

Methods:

wait_for_line(timeout=None)
Pause the script until the device is deactivated, or the timeout (in seconds) is reached.

wait_for_no_line(timeout=None)
Pause the script until the device is activated, or the timeout (in seconds) is reached.

when_line
The function to run when the device changes state from active to inactive.

when_no_line
The function to run when the device changes state from inactive to active.

pin
The GPIO pin that the device's output is connected to.

Motion Sensor

```
gpiozero.MotionSensor(pin, queue_len=1, sample_rate=10,
threshold=0.5, partial=False)
```

As shown in chapter 6, this class is used with a passive infrared (PIR) sensor, such as the one found in the CamJam EduKit #2. The following example will print a line of text when motion is detected by the sensor (with its middle output leg connected to GPIO pin 4):

```
from gpiozero import MotionSensor

pir = MotionSensor(4)
pir.wait_for_motion()
print("Motion detected!")
```

Methods:

wait_for_motion(timeout=None)
Pause the script until the device is activated, or the timeout (in seconds) is reached.

wait_for_no_motion(timeout=None)
Pause the script until the device is deactivated, or the timeout (in seconds) is reached.

motion_detected
Returns True if the device is currently active, and False otherwise.

when_motion
The function to run when the device changes state from inactive to active.

when_no_motion
The function to run when the device changes state from active to inactive.

pin
The GPIO pin that the device's output is connected to.

Light Sensor

```
gpiozero.LightSensor(pin, queue_len=5, charge_time_
limit=0.01, threshold=0.1, partial=False)
```

As shown in chapter 8. Connect one leg of the LDR to the 3V3 pin; connect one leg of a 1µF capacitor to a ground pin; connect the other leg of the LDR and the other leg of the capacitor to the same GPIO pin. This class repeatedly discharges the capacitor, then times the duration it takes to charge, which will vary according to the light falling on the LDR. The following code will print a line of text when light is detected:

```
from gpiozero import LightSensor

ldr = LightSensor(18)
ldr.wait_for_light()
print("Light detected!")
```

Methods:

wait_for_dark(timeout=None)
Pause the script until the device is deactivated, or the timeout (in seconds) is reached.

wait_for_light(`timeout=None`**)**
Pause the script until the device is activated, or the timeout (in seconds) is reached.

light_detected
Returns True if the device is currently active, and False otherwise.

when_dark
The function to run when the device changes state from active to inactive.

when_light
The function to run when the device changes state from inactive to active.

pin
The GPIO pin that the device is connected to.

Distance Sensor

```
gpiozero.DistanceSensor(echo, trigger, queue_len=30,
max_distance=1, threshold_distance=0.3, partial=False)
```

As shown in chapter 7, this class is used with a standard HC-SR04 ultrasonic distance sensor, as found in the CamJam EduKit #3. Note: to avoid damaging the Pi, you'll need to use a voltage divider on the breadboard to reduce the sensor's output (ECHO pin) from 5V to 3.3V. The following example will periodically report the distance measured by the sensor in cm (with the TRIG pin connected to GPIO17, and ECHO pin to GPIO18):

```
from gpiozero import DistanceSensor
from time import sleep

sensor = DistanceSensor(echo=18, trigger=17)
while True:
    print('Distance: ', sensor.distance * 100)
    sleep(1)
```

Methods:

distance
Returns the current distance measured by the sensor in metres.
Note that this property will have a value between 0 and **max_distance**.

max_distance
As specified in the class constructor, the maximum distance that
the sensor will measure in metres.

threshold_distance
As specified in the class constructor, the distance (measured in
metres) that will trigger the **when_in_range** and **when_out_of_range** events when crossed.

when_in_range
The function to run when the device changes state from active
to inactive.

when_out_of_range
The function to run when the device changes state from inactive
to active.

wait_for_in_range(timeout=None)
Pause the script until the device is deactivated, or the timeout
is reached.

wait_for_out_of_range(timeout=None)
Pause the script until the device is activated, or the timeout
is reached.

echo
Returns the GPIO pin that the sensor's ECHO pin is connected to.

trigger
Returns the GPIO pin that the sensor's TRIG pin is connected to.

03. SPI DEVICES

SPI (serial peripheral interface) is a mechanism allowing compatible devices to communicate with the Pi. GPIO Zero provides some classes for devices, including a range of analogue-to-digital converters...

[SPI ARGUMENTS]

When constructing an SPI device, there are two schemes for specifying which pins it's connected to...

1. You can specify port and device keyword arguments. The port parameter must be 0; there's only one user-accessible hardware SPI interface on the Pi, using GPIO11 as the clock pin, GPIO10 as the MOSI pin, and GPIO9 as the MISO pin. The device parameter must be 0 or 1. If device is 0, the select pin will be GPIO8; if device is 1, the select pin will be GPIO7.

2. Alternatively, you can specify clock_pin, mosi_pin, miso_pin, and select_pin keyword arguments. In this case, the pins can be any four GPIO pins. Remember that SPI devices can share clock, MOSI, and MISO pins, but not select pins; the GPIO Zero library will enforce this restriction.

   You can't mix these two schemes, but you can omit any arguments from either scheme. The defaults are:

> port and device both default to 0.

> clock_pin defaults to 11, mosi_pin defaults to 10, miso_pin defaults to 9, and select_pin defaults to 8.

Analogue-to-Digital Converters (ADCs)

MCP3001:
```
gpiozero.MCP3001(**spi_args)
```

MCP3002:
```
gpiozero.MCP3002(channel=0, differential=False, **spi_args)
```

MCP3004:
```
gpiozero.MCP3004(channel=0, differential=False, **spi_args)
```

MCP3008:
```
gpiozero.MCP3008(channel=0, differential=False, **spi_args)
```

MCP3201:
```
gpiozero.MCP3201(**spi_args)
```

MCP3202:
```
gpiozero.MCP3202(channel=0, differential=False, **spi_args)
```

MCP3204:
```
gpiozero.MCP3204(channel=0, differential=False, **spi_args)
```

MCP3208:
```
gpiozero.MCP3208(channel=0, differential=False, **spi_args)
```

MCP3301:
```
gpiozero.MCP3301(**spi_args)
```

MCP3302:
```
gpiozero.MCP3302(channel=0, differential=False, **spi_args)
```

MCP3304:
```
gpiozero.MCP3304(channel=0, differential=False, **spi_args)
```

GPIO Zero supports a range of ADC chips, with varying numbers of bits (from 10-bit to 13-bit) and channels (1 to 8). As shown in chapters 9 and 10, numerous jumper wires are required to connect the ADC via a breadboard to the Pi.

Methods:

channel
The channel to read data from. The MCP3008/3208/3304 have eight channels (0-7), while the MCP3004/3204/3302 have four channels (0-3), and the MCP3001/3201 only have one channel. The MCP3301 always operates in differential mode between its two channels, and the output value is scaled from -1 to +1.

differential
If True, the device is operated in pseudo-differential mode. In this mode, one channel (specified by the channel attribute) is read relative to the value of a second channel, informed by the chip's design.

value
The current value read from the device, scaled to a value between 0 and 1 (or -1 to +1 for devices operating in differential mode).

04. BOARDS & ACCESSORIES

To make things even easier, GPIO Zero provides extra support for a range of add-on devices and component collections...

LEDBoard

```
gpiozero.LEDBoard(*pins, pwm=False, active_high=True,
initial_value=False, **named_pins)
```

This class enables you to control a generic LED board or collection of LEDs. The following example turns on all the LEDs on a board containing five LEDs attached to GPIO pins 2 through 6:

```
from gpiozero import LEDBoard

leds = LEDBoard(2, 3, 4, 5, 6)
leds.on()
```

Methods:

on(*args)
Turn all the output devices on.

off(*args)
Turn all the output devices off.

blink(on_time=1, off_time=1, fade_in_time=0, fade_out_time=0, n=None, background=True**)**
Make all the LEDs turn on and off repeatedly.

pulse(fade_in_time=1, fade_out_time=1, n=None, background=True**)**
Make the device fade in and out repeatedly.

toggle(*args)
Toggle all the output devices. For each device, if it's on, turn it off; if it's off, turn it on.

leds
A flat tuple of all LEDs contained in this collection (and all sub-collections).

source
The iterable to use as a source of values for **value**.

source_delay
The delay (measured in seconds) in the loop used to read values from **source**. Defaults to 0.01 seconds.

value
A tuple containing a value for each subordinate device. This property can also be set to update the state of all subordinate output devices.

values
An infinite iterator of values read from **value**.

LEDBarGraph

```
gpiozero.LEDBarGraph(*pins, initial_value=0)
```

As shown in chapter 10, this is a class for controlling a line of LEDs to represent a bar graph. Positive values (0 to 1) light the LEDs from first to last. Negative values (-1 to 0) light the LEDs from last to first. The following example demonstrates turning on the first two and last two LEDs in a board containing five LEDs attached to GPIOs 2 through 6:

```
from gpiozero import LEDBarGraph
from time import sleep

graph = LEDBarGraph(2, 3, 4, 5, 6)
graph.value = 2/5  # Light the first two LEDs only
sleep(1)
graph.value = -2/5 # Light the last two LEDs only
sleep(1)
graph.off()
```

Methods:

on()
Turn all the output devices on.

off()
Turn all the output devices off.

toggle()
Toggle all the output devices. For each device, if it's on, turn it off; if it's off, turn it on.

leds
A flat tuple of all LEDs contained in this collection (and all sub-collections).

source
The iterable to use as a source of values for **value**.

source_delay
The delay (measured in seconds) in the loop used to read values from **source**. Defaults to 0.01 seconds.

value
A tuple containing a value for each subordinate device. This property can also be set to update the state of all subordinate output devices. To light a particular number of LEDs, simply divide that number by the total number of LEDs.

values
An infinite iterator of values read from **value**.

TrafficLights

```
gpiozero.TrafficLights(red=None, amber=None, green=None,
pwm=False, initial_value=False)
```

Extends **LEDBoard** for devices containing red, amber, and green LEDs (or individual LEDs). The following example initialises a device connected to GPIO pins 2, 3, and 4, then lights the amber LED attached to GPIO 3:

```
from gpiozero import TrafficLights

traffic = TrafficLights(2, 3, 4)
traffic.amber.on()
```

Methods:

These are the same as for the **LEDBoard** class.

PiLITEr

```
gpiozero.PiLiter(pwm=False, initial_value=False)
```

Extends **LEDBoard** for the Ciseco Pi-LITEr, a strip of eight very bright LEDs. The Pi-LITEr pins are fixed and therefore there's no need to specify them. The following example turns on all the LEDs of the Pi-LITEr:

```
from gpiozero import PiLiter

lite = PiLiter()
lite.on()
```

Methods:

These are the same as for the **LEDBoard** class.

SnowPi

```
gpiozero.SnowPi(pwm=False, initial_value=False)
```

Extends **LEDBoard** for the Ryanteck SnowPi board. Since its pins are fixed, there's no need to specify them. The following example turns on the eyes, sets the nose pulsing, and the arms blinking:

```
from gpiozero import SnowPi

snowman = SnowPi(pwm=True)
snowman.eyes.on()
snowman.nose.pulse()
snowman.arms.blink()
```

Methods:

These are the same as for the **LEDBoard** class.

PiLITEr Bar Graph

```
gpiozero.PiLiterBarGraph(pwm=False, initial_value=0.0)
```

Extends **LEDBarGraph** to treat the Ciseco Pi-LITEr as an eight-segment bar graph. The Pi-LITEr pins are fixed and therefore there's no need to specify them. The following example sets the graph value to 0.5:

```
from gpiozero import PiLiterBarGraph

graph = PiLiterBarGraph()
graph.value = 0.5
```

> ## Methods:
>
> These are the same as for the **LEDBarGraph** class.

PI-TRAFFIC

gpiozero.PiTraffic(pwm=False, initial_value=False)

Extends **TrafficLights** for the Low Voltage Labs Pi-Traffic, a vertical traffic lights board, when attached to GPIO pins 9, 10, and 11. The following example turns on the amber LED:

```
from gpiozero import PiTraffic

traffic = PiTraffic()
traffic.amber.on()
```

> ## Methods:
>
> These are the same as for the **TrafficLights** class.

TrafficLightsBuzzer

gpiozero.TrafficLightsBuzzer(lights, buzzer, button)

A generic class for HATs with traffic lights, a button, and a buzzer.

Methods:

on()
Turn all the output devices on.

off()
Turn all the output devices off.

toggle()
Toggle all the output devices. For each device, if it's on, turn it off;
if it's off, turn it on.

source
The iterable to use as a source of values for **value**.

source_delay
The delay (measured in seconds) in the loop used to read values from **source**.
Defaults to 0.01 seconds.

value
A tuple containing a value for each subordinate device.
This property can also be set to update the state of all subordinate output devices.

values
An infinite iterator of values read from **value**.

Fish Dish

gpiozero.FishDish(pwm=False)

Extends **TrafficLightsBuzzer** for the Pi Supply Fish Dish, which has traffic light LEDs,
a button, and a buzzer. Since its pins are fixed, there's no need to specify them. The following
example waits for the button to be pressed on the Fish Dish, then turns on all the LEDs:

```
from gpiozero import FishDish

fish = FishDish()
```

```
fish.button.wait_for_press()
fish.lights.on()
```

Methods:

These are the same as for the **TrafficLightsBuzzer** class.

Traffic HAT

`gpiozero.TrafficHat(pwm=False)`

Extends **TrafficLightsBuzzer** for the Ryanteck Traffic HAT, which has traffic light LEDs, a button, and a buzzer. Since its pins are fixed, there's no need to specify them. The following example waits for the button to be pressed on the Traffic HAT, then turns on all the LEDs:

```
from gpiozero import TrafficHat

hat = TrafficHat()
hat.button.wait_for_press()
hat.lights.on()
```

Methods:

These are the same as for the **TrafficLightsBuzzer** class.

Robot

`gpiozero.Robot(left=None, right=None)`

Designed to control a generic dual-motor robot (as in chapter 11), this class is constructed with two tuples representing the forward and backward pins of the left and right controllers respectively. For example, if the left motor's controller is connected to GPIOs 4 and 14, while the right motor's controller is connected to GPIOs 17 and 18, then the following example will drive the robot forward:

```
from gpiozero import Robot

robot = Robot(left=(4, 14), right=(17, 18))
robot.forward()
```

Methods:

forward(speed=1**)**
Drive the robot forward by running both motors forward.

backward(speed=1**)**
Drive the robot backward by running both motors backward.

left(speed=1**)**
Make the robot turn left by running the right motor forward and left motor backward.

right(speed=1**)**
Make the robot turn right by running the left motor forward and right motor backward.

reverse()
Reverse the robot's current motor directions. If the robot is currently running full speed forward, it will run full speed backward. If the robot is turning left at half speed, it will turn right at half speed. If the robot is currently stopped, it will remain stopped.

stop()
Stop the robot.

source
The iterable to use as a source of values for **value**.

source_delay
The delay (measured in seconds) in the loop used to read values from **source**. Defaults to 0.01 seconds.

value
Represents the motion of the robot as a tuple of (**left_motor_speed**, **right_motor_speed**), with (-1, -1) representing full speed backward, (1, 1) representing full speed forward, and (0, 0) representing stopped.

> **values**
> An infinite iterator of values read from **value**.

Ryanteck MCB Robot

gpiozero.RyanteckRobot

Extends **Robot** for the Ryanteck MCB robot. Since its pins are fixed, there's no need to specify them when constructing this class. The following example drives the robot forward:

```
from gpiozero import RyanteckRobot

robot = RyanteckRobot()
robot.forward()
```

> **Methods:**
>
> These are the same as for the **Robot** class.

CamJam #3 Kit Robot

gpiozero.CamJamKitRobot

Extends **Robot** for the CamJam EduKit #3 robot. Since its pins are fixed, there's no need to specify them when constructing this class. The following example drives the robot forward:

```
from gpiozero import CamJamKitRobot

robot = CamJamKitRobot()
robot.forward()
```

> **Methods:**
>
> These are the same as for the **Robot** class.